LER 2562

AF711749

ADDING & SUBTRACTING FRACTIONS WITH LIKE DENOMINATORS

by
Barbara Bando Irvin, Ph.D.

TABLE OF CONTENTS

INTRODUCTION

Adding & Subtracting With Like Denominators is included in a series of 32-page activity books developed to help students in grades 3-6 understand, compare, and perform operations on fractions. The activities in each book provide hands-on experiences using Fraction Tower® Cubes with 24 blackline master activity sheets. Students can build fraction towers, record their discoveries, and complete various exercises on these activity pages. Refer to the scope and sequence chart on the inside front cover for the content of each book, and for suggested grade levels.

The Teaching Notes on pages 4-6 offer explanations and suggestions to help students understand the activities. Selected Solutions on page 32 provide at least one solution for each problem. On the inside back cover, a Glossary is provided for quick reference of fraction terms. Students can make their own paper models of the Fraction Tower Cubes using copies of the Fraction Tower Chart on page 7. They can also use the Fraction Tower Chart as a workmat to build and compare fraction towers.

NCTM Standards

The fraction activities were developed using NCTM's *Curriculum and Evaluation Standards for School Mathematics* (1989) as a guide. Special attention was given to Standard 12 in grades K-4 and Standards 5, 6, and 7 in grades 5-8. They include the following goals for students:

- Develop concepts for fractions and mixed numbers.
- Develop and use order relationships for fractions.
- Develop number sense for fractions.
- Use models to find equivalent fractions.
- Use models to explore operations on fractions.

About the Fraction Tower Cubes

The set of Fraction Tower Cubes consists of 51 interlocking, proportional-sized cubes representing unit fractions $\frac{1}{12}$, $\frac{1}{10}$, $\frac{1}{8}$, $\frac{1}{6}$, $\frac{1}{5}$, $\frac{1}{4}$, $\frac{1}{3}$, $\frac{1}{2}$, and the number 1. They are color-coded in nine colors to match other Learning Resources fraction manipulatives: Deluxe Rainbow Fraction Circles, Deluxe Rainbow Fraction Squares, and Rainbow Fraction Tiles. This system enables students to make connections among different-shaped fractional parts as they work with different fraction models.

The set of 51 Fraction Tower Cubes includes:

- one red *whole*
- two pink *halves*
- three orange *thirds*
- four yellow *fourths*
- five green *fifths*
- six teal (blue-green) *sixths*
- eight blue *eighths*
- ten purple *tenths*
- twelve black *twelfths*

Unlike most other fraction manipulatives, the Fraction Tower Cubes are three-dimensional and can be snapped together to create a single piece. For example, five blue cubes can be snapped together to show the fraction $\frac{5}{8}$. Then the tower can be handled as one piece to compare it to another fraction, such as $\frac{3}{4}$, represented by snapping together three yellow cubes.

Free Exploration

Provide time for students to familiarize themselves with the Fraction Tower Cubes before doing any structured activities. Allow them to snap the cubes together and then unsnap the cubes to get a "feel" for them. Observe and listen to the students' comments. Ask some informal questions to assess their understanding of fractions.

Classroom Management

Students can work individually or in pairs to complete each activity. As students clean up after an activity, have them snap together the cubes to make same-color fraction towers, and place them back into the self-closing plastic containers. This is an easy way to check if any cubes are missing, and the set of Fraction Tower Cubes is ready to be used again.

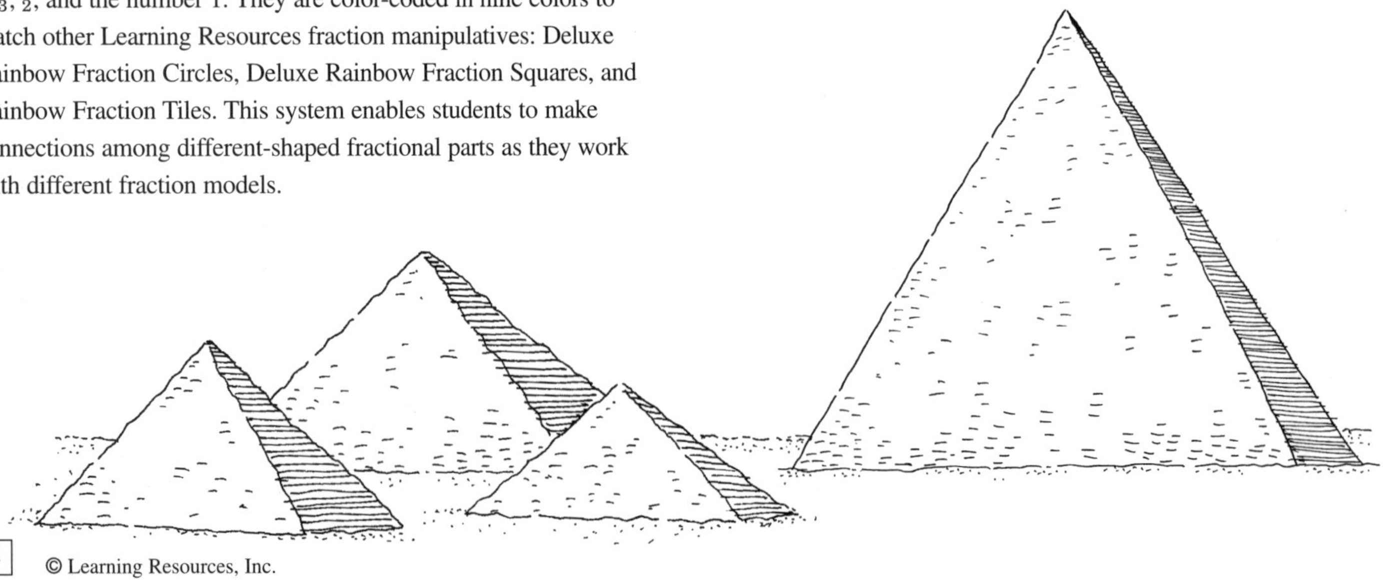

HOW TO USE THE ACTIVITIES

It is very important that students have adequate time to understand the size of fractional numbers and the relationships between them. All too often, these types of activities with fractions are rushed or cut short in order to focus on the algorithms for performing the operations on fractions. The activities in this book center around adding and subtracting fractions and mixed numbers with like denominators. A great deal of attention is given to joining (snapping) and separating (unsnapping) the cubes to understand these two operations. Use the activities to introduce, enhance, or even replace a classroom lesson. Discuss the students' ideas and solutions.

Workmat Page Format

The pages in this activity book resemble "workmats" so students can physically manipulate fraction cubes in an adequate amount of space. The mats are designed to align the bottoms of the fraction towers in order to build and compare fractions. The red cube is often displayed on the left side of the page to compare fraction towers to 1. These workmat pages help students connect actual fraction towers to pictorial images of them, and then to the written fraction. Horizontal and vertical formats help students draw and write their responses directly on the activity page without having to use another sheet of paper.

Many of the activities are arranged in a two-page layout. The pages may be duplicated for classroom use, or books may be purchased in sufficient numbers for each student to have his or her own. The Fraction Tower Chart on page 7 can be used to make a set of Fraction Tower Cube cutouts for students who want their own set of fraction cubes. A transparency of the chart for the overhead projector can be used to build and compare fractions, improper fractions, and mixed numbers.

Materials Needed

The activities in this book require the following materials:

- two sets of Fraction Tower® Cubes
- a box of crayons (colored pencils, markers)
- a pencil
- paper clip or clear spinner
- game markers (coins, erasers, buttons, etc.)

Warm Up

The activities in Activity Books 1 and 2 focused on building and comparing fractions and mixed numbers. Before starting the activities in this book, do a quick review by having students complete the following tasks:

1. Review how to build fraction towers. Build fraction towers for fractions such as $\frac{5}{6}$, $\frac{5}{12}$, or $\frac{7}{10}$. How do you know how to build a fraction tower for a fraction? *(Look at the numerator and denominator. To build $\frac{5}{6}$, use 5 teal cubes.)*

2. Review equivalent fractions. Build some one-color fraction towers that are as tall as $\frac{3}{4}$. *($\frac{3}{4}$: 3 yellow cubes; $\frac{6}{8}$: 6 blue cubes; $\frac{9}{12}$: 9 black cubes)*

3. How do you know you found the simplest form of a fraction when you build a fraction tower? For fractions $\frac{4}{6}$ and $\frac{2}{3}$, which fraction is in simplest form? Why? *(A fraction tower for a fraction in simplest form uses the fewest number of one-color cubes. For $\frac{4}{6}$, use 4 teal cubes. For $\frac{2}{3}$, use only 2 orange cubes.)*

4. Build the mixed number $1\frac{3}{10}$. *(1 red cube and 3 purple cubes.)* How can you express this mixed number as an improper fraction? *(Make a one-color fraction tower with 13 purple cubes.)*

5. Review comparing fractions and order. Build fraction towers for $\frac{5}{8}$, $\frac{8}{5}$, $1\frac{3}{8}$, and $\frac{5}{4}$. Put them in order from shortest to tallest. [$\frac{5}{8}$, $\frac{5}{4}$ ($1\frac{1}{4}$), $1\frac{3}{8}$, $\frac{8}{5}$ ($1\frac{3}{5}$)]

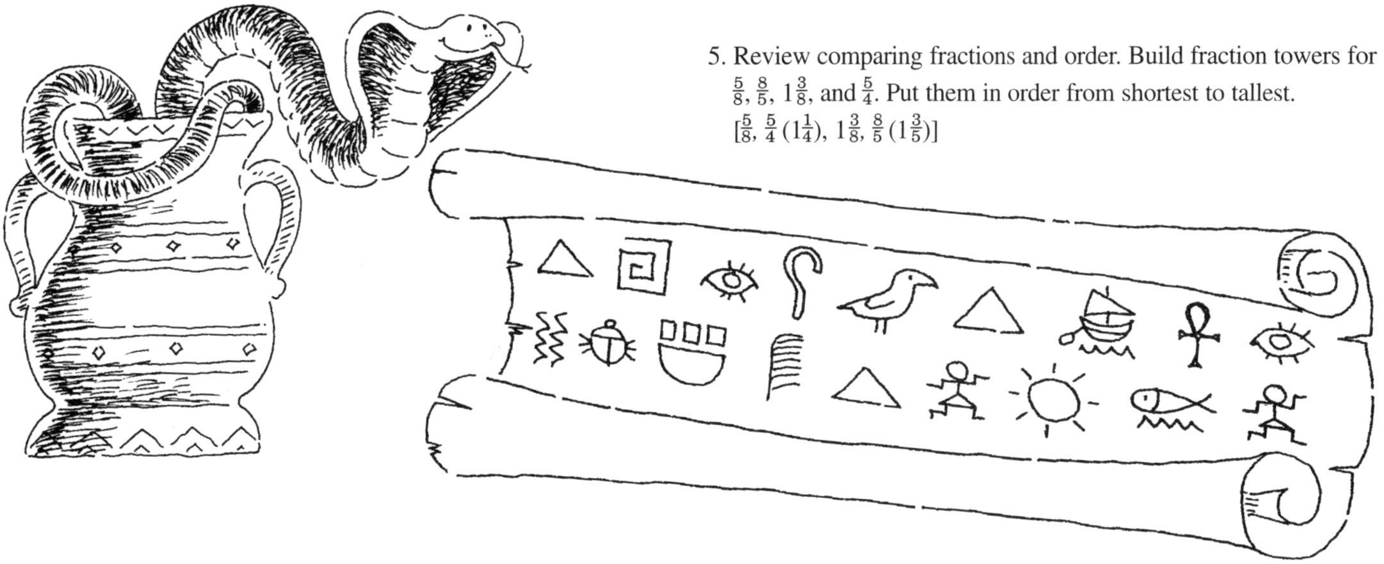

TEACHING NOTES

Note: The "build and draw" direction on the activity pages means to build a fraction with Fraction Tower® Cubes and to trace (or draw) the cubes on the activity page, remove the cubes, label each unit fraction part, and then color the fraction tower the appropriate color with crayons, colored pencils, or markers. To color the black cubes, students may wish to shade them with a pencil so as not to color over the labeling.

The students only need one set of Fraction Tower Cubes for pages 8-11.

Adding Fractions (pages 8-9)

The process of building the two addends and then snapping them together to find the sum is shown on page 8. After students have completed this page, talk about the two problems and have students demonstrate the addition process. Also, explain the terms *addend* and *sum*.

Seven problems are provided on page 9. The grid at the right helps students build fraction towers to find the sum. Students may realize that when they add fractions with like denominators (fraction tower addends in the same color), all they have to do is add the numerators. The last problem shows that more than two fractions may be added to find a sum.

Adding Fractions and Simplifying Sums (pages 10-11)

Another step is added to the process of adding fractions in this lesson. The sum must be given in simplest form. If some students have forgotten how to find an equivalent fraction that uses the fewest number of cubes for a fraction, plan a short review before doing this page.

On page 11, there are two answer blanks for each problem, one for the sum, and one for the sum in simplest form.

The students will need two sets of Fraction Tower Cubes for pages 12-17.

Adding Fractions and Mixed Number Sums (pages 12-13)

In this lesson, the sums are greater than one. After students discover that the sums on these pages are improper fractions, they must change them to mixed numbers. A fraction tower for an improper fraction is a one-color tower equal to or greater than 1 red tower. A mixed number can be shown using a red tower for 1 and a second color of cubes for the fraction part. Fraction tower mixed numbers can be displayed as one tall, two-color fraction tower using red and other color cubes, or as one red and a second color fraction tower.

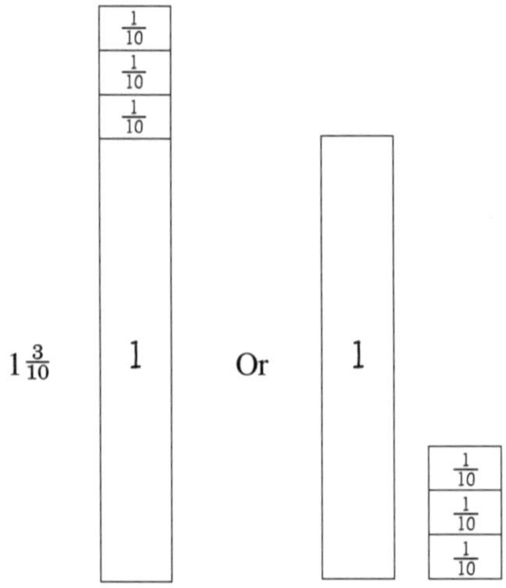

On page 13, there are two answer blanks for each problem, one for the improper fraction sum, and one for the mixed number sum.

Simplifying Large Sums (pages 14-15)

This lesson is the same as the previous one except that the sums must now be expressed in simplest form. Watch students as they build and draw the addends and sum on page 14. Three answer blanks are provided for each problem on page 15. The first one is for the sum (improper fraction), the second one is for the mixed number sum, and the third one is for the mixed number sum in simplest form.

Fraction and Mixed Number Sums (page 16)

After finding sums less than, equal to, and greater than 1, ask students to predict whether a sum will be less than or greater than 1 for an addition problem. Have them look at the addends of each problem on this page and circle the problems with sums less than 1, and ask them for their reasons. The set of addition problems on this page is a review of those on pages 8-15.

TEACHING NOTES

Missing Addends (page 17)

Present a problem like $\frac{5}{8}$ + ___ = $\frac{3}{4}$ to the students and ask them how to find the missing addend. Since the sum $\frac{3}{4}$ has a different denominator than the addend $\frac{5}{8}$, then it must be in simplest form. Have them build a fraction tower for the sum (3 yellow cubes), and then find an equivalent fraction tower for eighths (6 blue for $\frac{6}{8}$) to consider the problem $\frac{5}{8}$ + ___ = $\frac{6}{8}$. Now that the addend and the sum have like denominators, it's easy to see that $\frac{1}{8}$ is the missing addend. If the sum is a mixed number, have them build a one-color improper fraction with the same denominator as the addend.

The students need only one set of Fraction Tower® Cubes for pages 18-23.

Subtracting Fractions (pages 18-19)

Subtraction can be thought of in two ways: *take away* and *comparison*. On pages 18 and 19, the "take away" technique is used primarily because it takes fewer cubes to show the subtraction process. When subtracting fractions with like denominators, build a fraction tower for the first number (minuend), and then simply unsnap the number of cubes to correspond to the numerator in the second number (subtrahend). "What's left?" is the difference.

If you wish to show students how to subtract using the "comparison" technique, have them build fraction towers for both numbers (minuend and subtrahend) in the subtraction problem. Then place them side by side to compare them to see the difference in height between the two towers.

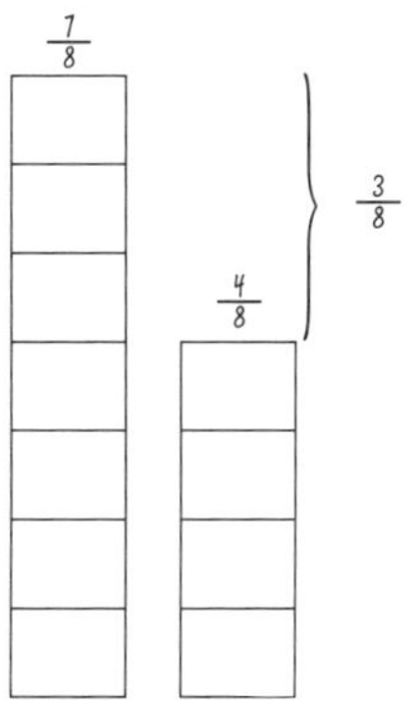

$\frac{7}{8}$ is taller than $\frac{4}{8}$

The difference between the heights is $\frac{3}{8}$.

Subtracting Fractions, Simplifying Differences (pages 20-21)

After students find the difference for a problem, they must express it in the simplest form. There are two answer blanks for each problem on page 23. One is for the difference, and one is for the difference in simplest form. For problem 6, the difference is in simplest form.

Subtracting Fractions From One (pages 22-23)

For this lesson, students must know how to express 1 as an improper fraction. A review is provided on page 22 to show that $1 = \frac{2}{2} = \frac{3}{3} = \frac{4}{4}$ and so on.

To subtract a fraction from 1, the 1 must be expressed as an improper fraction with the same denominator as the fraction being subtracted. To subtract $1 - \frac{4}{5}$, use $\frac{5}{5}$ for 1 ($\frac{5}{5} - \frac{4}{5}$). Remind students to express the differences on page 23 in simplest form.

The students will need two sets of Fraction Tower Cubes for pages 24-31.

Subtracting From Mixed Numbers (pages 24-25)

Although subtracting a fraction from a mixed number is a new concept for students, the process is still the same: subtract the fraction parts, and then subtract the whole-number parts. Finally, express the difference in simplest form. Problem 7 on page 25 is not as simple as the others, and is meant to be used as a prelude to the problems on pages 26 and 27. This problem involves regrouping the mixed number before any subtracting can be done. Ask students how they would solve this problem. If they have difficulty, save it to use as an example to begin the next lesson.

Subtracting With Regrouping (pages 26-27)

For $1\frac{3}{8} - \frac{5}{8}$, $\frac{5}{8}$ cannot be taken away from $\frac{3}{8}$. However, $\frac{5}{8}$ can be taken away from the mixed number $1\frac{3}{8}$ because $1\frac{3}{8} > \frac{5}{8}$. To make this physically possible, $1\frac{3}{8}$ must be changed to have smaller parts (eighths). Since $1\frac{3}{8} = 1 + \frac{3}{8}$, then $1 + \frac{3}{8} = \frac{8}{8} + \frac{3}{8} = \frac{11}{8}$. Now 5 eighths can physically be taken away from 11 eighths. Thus, $\frac{11}{8} - \frac{5}{8} = \frac{6}{8}$ and $\frac{6}{8} = \frac{3}{4}$ in simplest form.

Help students who have difficulty. Have them show you how to find the differences to at least two problems on page 27.

TEACHING NOTES

Subtracting Fractions and Mixed Numbers (page 28)

The problems on this page can be used as a review or anassessment tool. Allow students to use the Fraction Tower® Cubes to find each difference. Remind them to express the differences in simplest form.

Adding or Subtracting (page 29)

The problems on this page also can be used as a review or an assessment tool. Interview the students individually to see if they remember how to add fractions and can use the Fraction Tower Cubes to explain their answers. Make sure they pay attention to the addition and subtraction symbols in the problems.

Make-A-One Game (page 30)

This game provides an enjoyable way to practice adding fractions with like denominators. When each player chooses his or her color, one-color fraction towers are built to equal the sum of 1.

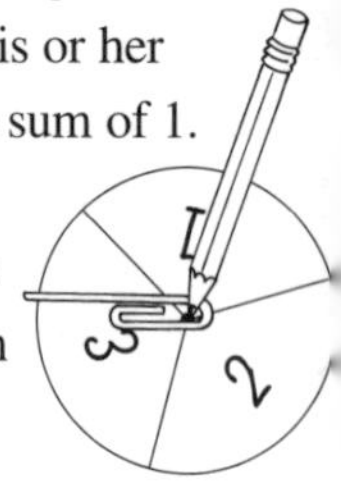

Modify the game by having a player say or write the addition number sentence for his or her tower at each turn. For example, if a player has 3 blue cubes and spins a 2, the player can say "$\frac{3}{8}$ plus $\frac{2}{8}$ equals $\frac{5}{8}$" or write "$\frac{3}{8} + \frac{2}{8} = \frac{5}{8}$."

A paper clip and pencil can be used on the spinner on the page.

Spin-A-Sum Game (page 31)

This game is a little more challenging than the Make-A-One Game. Students review adding fractions and must express sums in simplest form. In step 4, a third player (or you) can make sure the sum is correct and in simplest form. The markers that players use can be any small objects such as beans, jacks, or rocks.

FRACTION TOWER CHART

red	pink	yellow	blue	orange	teal	black	green	purple
1	$\frac{1}{2}$	$\frac{1}{4}$	$\frac{1}{8}$	$\frac{1}{3}$	$\frac{1}{6}$	$\frac{1}{12}$	$\frac{1}{5}$	$\frac{1}{10}$
1	$\frac{1}{2}$	$\frac{1}{4}$	$\frac{1}{8}$	$\frac{1}{3}$	$\frac{1}{6}$	$\frac{1}{12}$	$\frac{1}{5}$	$\frac{1}{10}$
	$\frac{1}{2}$	$\frac{1}{4}$	$\frac{1}{8}$	$\frac{1}{3}$	$\frac{1}{6}$	$\frac{1}{12}$	$\frac{1}{5}$	$\frac{1}{10}$
	$\frac{1}{2}$	$\frac{1}{4}$	$\frac{1}{8}$	$\frac{1}{3}$	$\frac{1}{6}$	$\frac{1}{12}$	$\frac{1}{5}$	$\frac{1}{10}$
		$\frac{1}{4}$	$\frac{1}{8}$	$\frac{1}{3}$	$\frac{1}{6}$	$\frac{1}{12}$	$\frac{1}{5}$	$\frac{1}{10}$
		$\frac{1}{4}$	$\frac{1}{8}$	$\frac{1}{3}$	$\frac{1}{6}$	$\frac{1}{12}$	$\frac{1}{5}$	$\frac{1}{10}$
		$\frac{1}{4}$	$\frac{1}{8}$		$\frac{1}{6}$	$\frac{1}{12}$	$\frac{1}{5}$	$\frac{1}{10}$
		$\frac{1}{4}$	$\frac{1}{8}$		$\frac{1}{6}$	$\frac{1}{12}$	$\frac{1}{5}$	$\frac{1}{10}$
			$\frac{1}{8}$		$\frac{1}{6}$	$\frac{1}{12}$	$\frac{1}{5}$	$\frac{1}{10}$
			$\frac{1}{8}$		$\frac{1}{6}$	$\frac{1}{12}$	$\frac{1}{5}$	$\frac{1}{10}$
			$\frac{1}{8}$		$\frac{1}{6}$	$\frac{1}{12}$		$\frac{1}{10}$
			$\frac{1}{8}$		$\frac{1}{6}$	$\frac{1}{12}$		$\frac{1}{10}$
			$\frac{1}{8}$			$\frac{1}{12}$		$\frac{1}{10}$
			$\frac{1}{8}$			$\frac{1}{12}$		$\frac{1}{10}$
			$\frac{1}{8}$			$\frac{1}{12}$		$\frac{1}{10}$
			$\frac{1}{8}$			$\frac{1}{12}$		$\frac{1}{10}$
						$\frac{1}{12}$		$\frac{1}{10}$
						$\frac{1}{12}$		$\frac{1}{10}$
						$\frac{1}{12}$		$\frac{1}{10}$
						$\frac{1}{12}$		$\frac{1}{10}$
						$\frac{1}{12}$		
						$\frac{1}{12}$		
						$\frac{1}{12}$		
						$\frac{1}{12}$		

ADDING FRACTIONS

To add fractions, build a fraction tower for each addend and snap them together to find the sum. Build and draw the addends, and find the sums for problems 1 and 2.

Name ______________________

1. addend $\frac{3}{10}$ + addend $\frac{4}{10}$ = sum $\frac{\square}{10}$

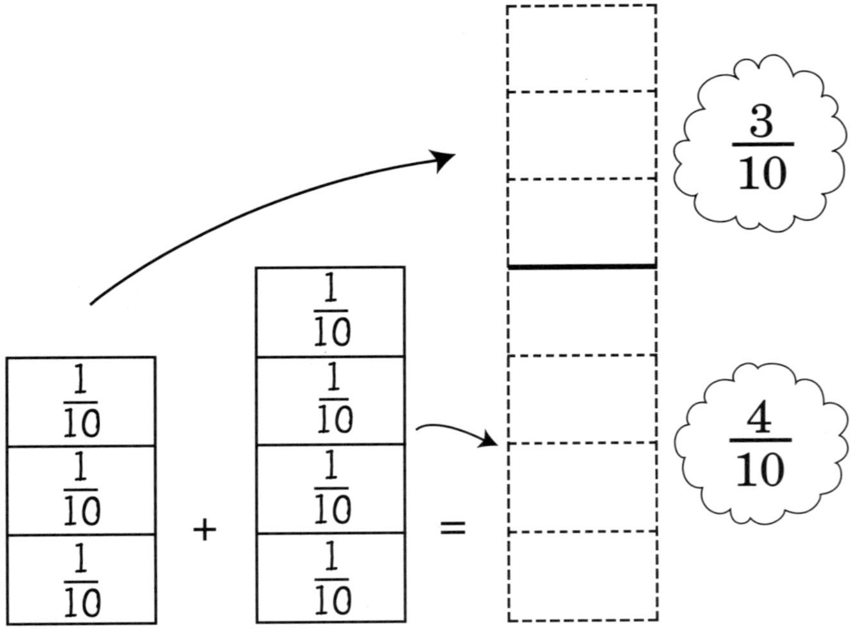

$\frac{3}{10} + \frac{4}{10} = \frac{\square}{10}$

_____ purple + _____ purple = _____ purple

2. addend $\frac{5}{8}$ + addend $\frac{2}{8}$ = sum $\frac{\square}{8}$

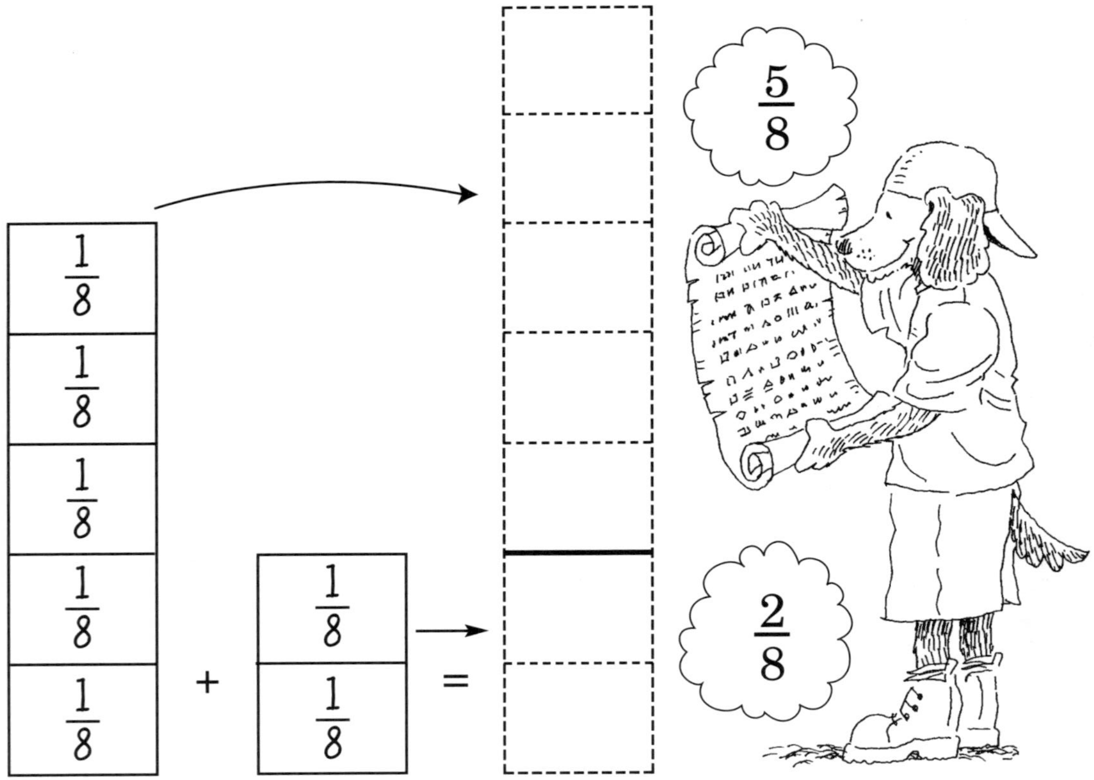

$\frac{5}{8} + \frac{2}{8} = \frac{\square}{8}$

_____ blue + _____ blue = _____ blue

ADDING MORE FRACTIONS

Use the grids below to build the addends and find the sum. Name ______________________

1. $\frac{2}{5} + \frac{2}{5} =$ _____

2. $\frac{1}{8} + \frac{6}{8} =$ _____

3. $\frac{4}{6} + \frac{1}{6} =$ _____

4. $\frac{2}{12} + \frac{3}{12} =$ _____

5. $\frac{2}{10} + \frac{5}{10} =$ _____

6. $\frac{8}{12} + \frac{2}{12} =$ _____

Try this one.

7. $\frac{4}{12} + \frac{2}{12} + \frac{5}{12} =$ _____

addend	+	addend	=	sum

ADDING FRACTIONS, SIMPLIFYING SUMS

Build and draw the addends to find the sums for problems 1 and 2. Name ______________________
Express the sum in simplest form.

1. addend, addend, sum, simplest form

$$\frac{2}{8} + \frac{4}{8} = \frac{\square}{8} = \frac{\square}{\square}$$

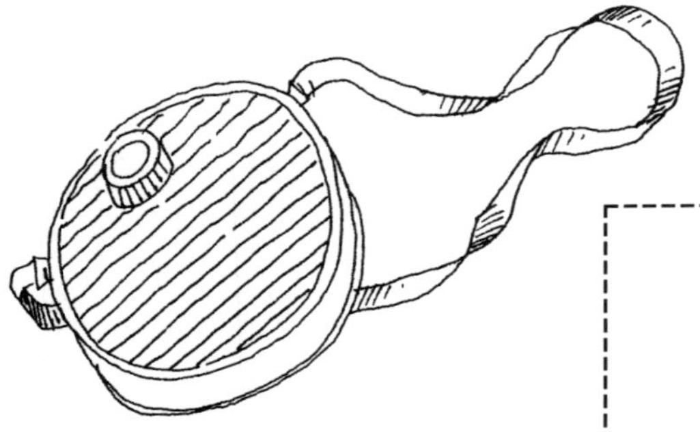

$\frac{1}{8}$ $\frac{1}{8}$ + ☐ = ☐ = ☐

$$\frac{2}{8} + \frac{4}{8} = \frac{\square}{8} = \frac{\square}{4}$$

_____ blue + _____ blue = _____ blue = _____ yellow

2. addend, addend, sum, simplest form

$$\frac{5}{12} + \frac{3}{12} = \frac{\square}{12} = \frac{\square}{\square}$$

☐ + ☐ = ☐ = ☐

$$\frac{5}{12} + \frac{3}{12} = \frac{\square}{12} = \frac{\square}{3}$$

_____ black + _____ black = _____ black = _____ orange

ADDING FRACTIONS AND SIMPLIFYING SUMS

Name ______________________

Use the grids below to build the addends, the sum, and then the simplest sum.

		sum		simplest sum
1.	$\frac{1}{4} + \frac{1}{4}$ =	____	=	____
2.	$\frac{2}{6} + \frac{1}{6}$ =	____	=	____
3.	$\frac{3}{12} + \frac{3}{12}$ =	____	=	____
4.	$\frac{6}{10} + \frac{2}{10}$ =	____	=	____
5.	$\frac{5}{12} + \frac{4}{12}$ =	____	=	____
6.	$\frac{4}{8} + \frac{2}{8}$ =	____	=	____

addends	+	addends	=	sum	=	simplest sum

ADDING FRACTIONS TO FIND MIXED NUMBER SUMS

Build and draw the addends, the sum, and then the mixed number sum. Name ______________________

addend		addend		sum		mixed number sum
$\frac{3}{6}$	+	$\frac{4}{6}$	=	$\frac{\square}{6}$	=	$\_\_\_\_ \frac{\square}{6}$

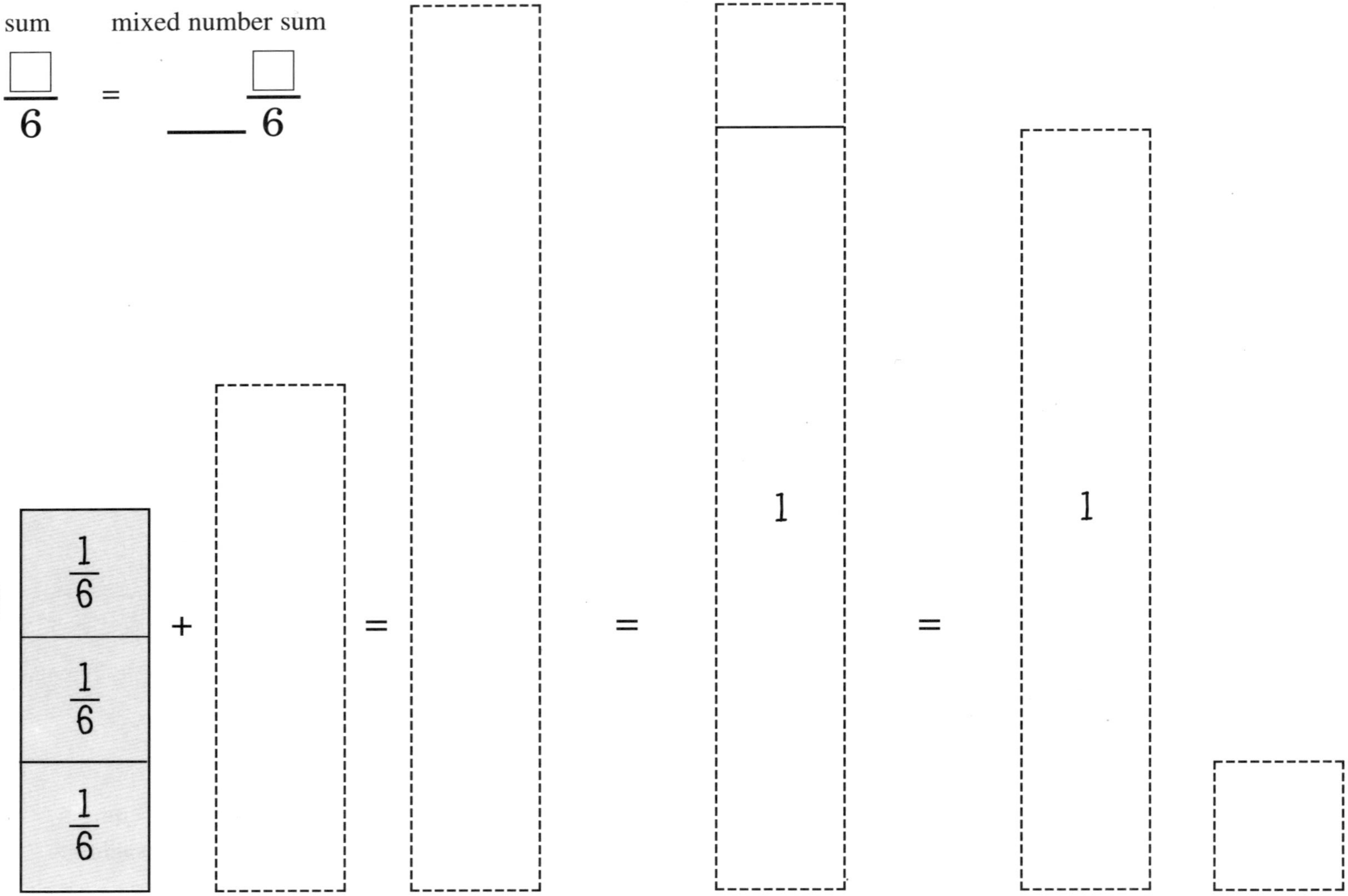

$$\frac{3}{6} + \frac{4}{6} = \frac{\square}{6} = \_\_\_\_ \frac{\square}{6} = \_\_\_\_ \frac{\square}{6}$$

_____ teal + _____ teal = _____ teal = 1 red, _____ teal = _____ red, _____ teal

ADDING FRACTIONS TO FIND MORE MIXED NUMBER SUMS

Use the grids below to build the addends, the sum, and then the mixed number sum.

Name ______________________

		sum	mixed number sum
1.	$\frac{2}{4} + \frac{3}{4}$	= ____	= ____
2.	$\frac{7}{10} + \frac{4}{10}$	= ____	= ____
3.	$\frac{4}{5} + \frac{4}{5}$	= ____	= ____
4.	$\frac{10}{12} + \frac{7}{12}$	= ____	= ____
5.	$\frac{4}{8} + \frac{5}{8}$	= ____	= ____
6.	$\frac{9}{12} + \frac{10}{12}$	= ____	= ____

+ = = 1

addend + addend = sum = mixed number sum

SIMPLIFYING LARGE SUMS

Build and draw the addends and the sum as an improper fraction, a mixed number, and finally as a mixed number in simplest form.

Name ______________________________

	improper fraction sum	mixed number sum	simplest mixed number sum

$$\frac{5}{6} + \frac{4}{6} = \frac{\square}{6} = \underline{\quad}\,\frac{\square}{6} = \underline{\quad}\,\frac{\square}{2}$$

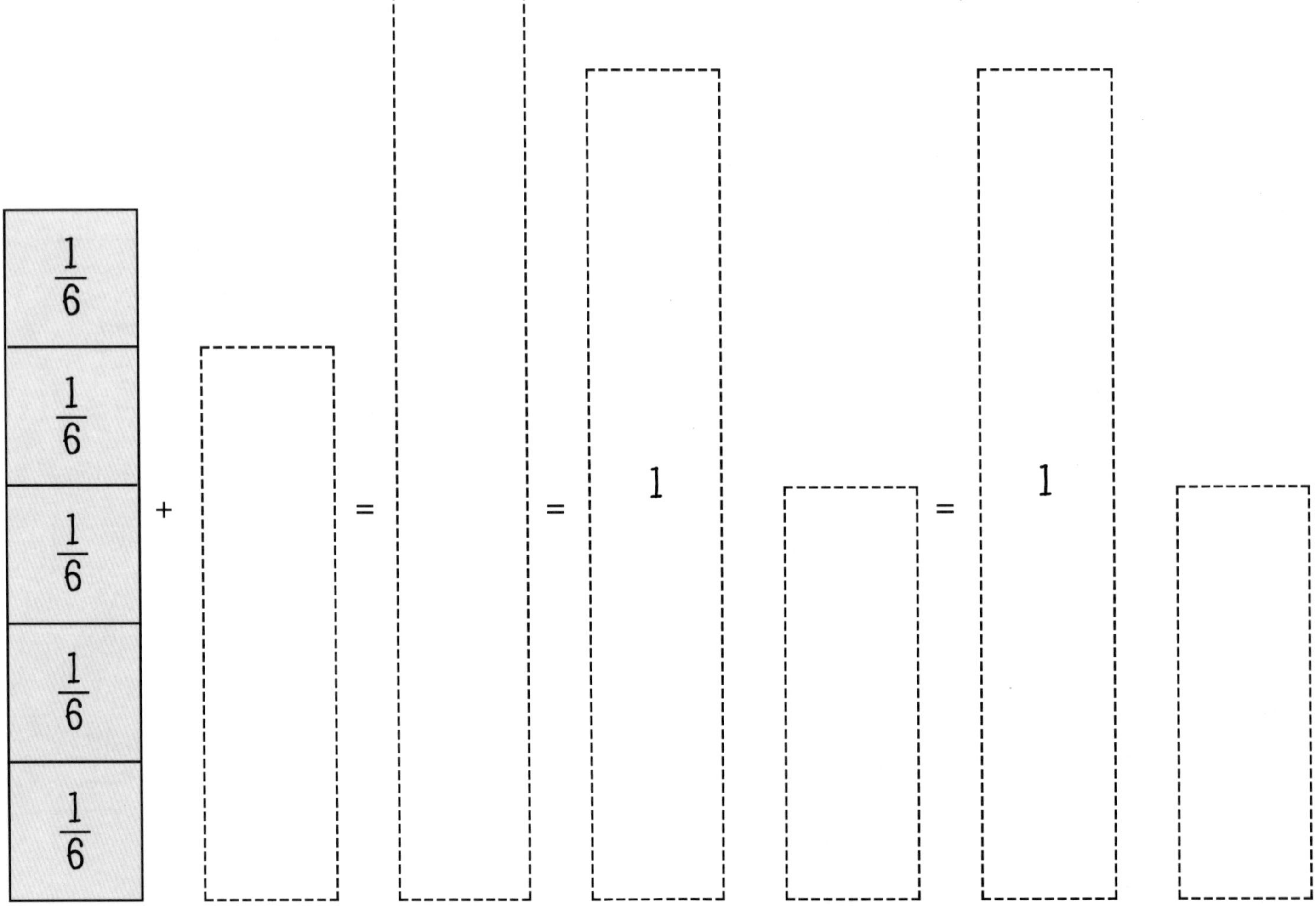

_____ teal + _____ teal = _____ teal = _____ red, _____ teal = _____ red, _____ pink

SIMPLIFYING MORE LARGE SUMS

Use the grids below to build addends, the sum, the mixed number sum, and the simplest mixed number sum.

Name ________________________________

	improper fraction sum	mixed number sum	simplest mixed number sum
1. $\frac{5}{8} + \frac{7}{8}$ =	____ =	____ =	____
2. $\frac{5}{6} + \frac{5}{6}$ =	____ =	____ =	____
3. $\frac{9}{12} + \frac{11}{12}$ =	____ =	____ =	____
4. $\frac{7}{10} + \frac{5}{10}$ =	____ =	____ =	____
5. $\frac{11}{12} + \frac{10}{12}$ =	____ =	____ =	____

addend + addend = improper fraction sum = mixed number sum = simplest mixed number sum

FRACTION AND MIXED NUMBER SUMS

Name ______________________

A. Guess which problems will have sums less than 1. Circle the problem numbers.
B. Use the grids below to build the addends and the sum. Write the sum in simplest form on the answer blank.

1. $\frac{3}{8} + \frac{7}{8} =$ ____

2. $\frac{7}{10} + \frac{2}{10} =$ ____

3. $\frac{5}{6} + \frac{5}{6} =$ ____

4. $\frac{5}{12} + \frac{3}{12} =$ ____

5. $\frac{6}{10} + \frac{7}{10} =$ ____

6. $\frac{7}{12} + \frac{9}{12} =$ ____

addend	+	addend	=	sum

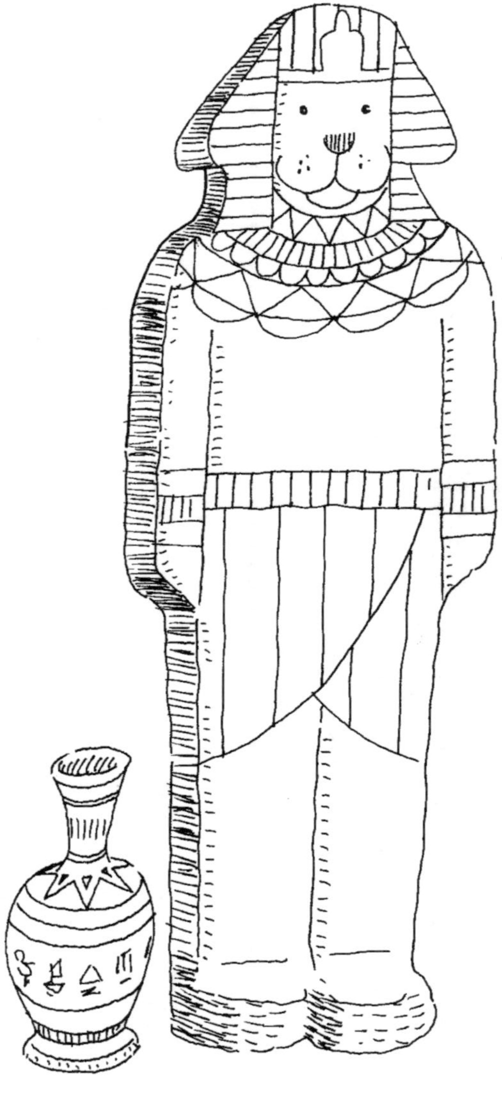

MISSING ADDENDS

Use the grids below to build the sum. If the sum is a one-color fraction tower in the color of the addend, unsnap the given addend part to see what is left. "What's left" is the missing addend.

Name ______________________

1. $\frac{6}{12} + \_\_\_\_ = \frac{12}{12}$

2. $\frac{2}{10} + \_\_\_\_ = \frac{1}{2}$

Think: $\frac{1}{2} = \frac{?}{10}$

3. $\frac{10}{12} + \_\_\_\_ = 1\frac{5}{12}$

4. $\_\_\_\_ + \frac{4}{8} = 1\frac{3}{8}$

5. $\frac{8}{12} + \_\_\_\_ = 1\frac{1}{4}$

6. $\_\_\_\_ + \frac{5}{6} = 1\frac{2}{3}$

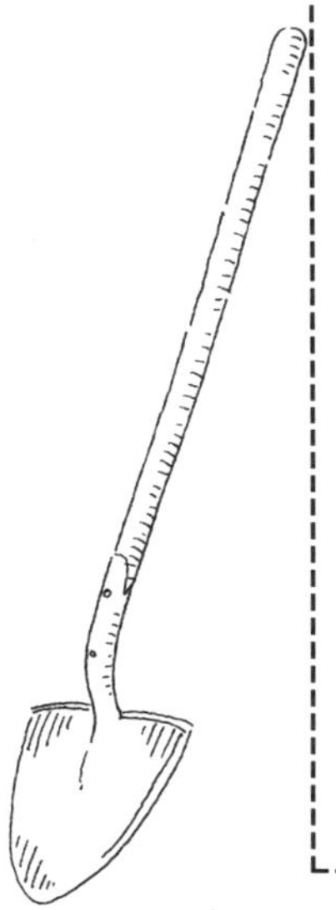

+ =

addend + addend = sum

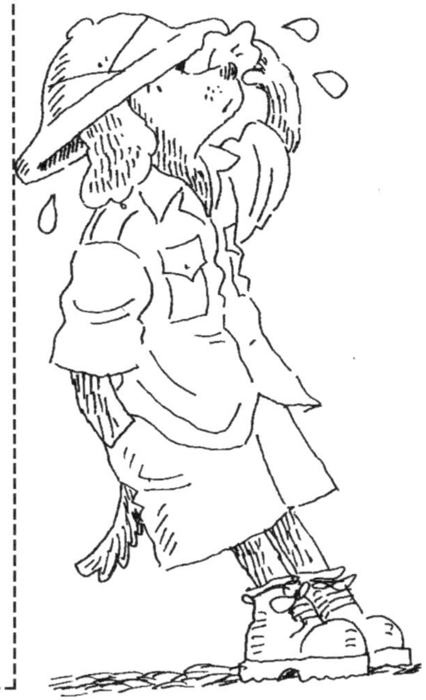

SUBTRACTING FRACTIONS

To subtract fractions, build the first number in the problem. Then "take away" the number of cubes in the second number to find the difference. Build and draw the problem and the difference.

Name ______________________

	Build this tower.		Take away this many.		Difference
1.	$\frac{7}{8}$	$-$	$\frac{4}{8}$	$=$	$\frac{\square}{8}$

7-cube tower
$\frac{1}{8}$
$\frac{1}{8}$
$\frac{1}{8}$
$\frac{1}{8}$
$\frac{1}{8}$
$\frac{1}{8}$
$\frac{1}{8}$

Take away $\frac{4}{8}$

What's left? →

What's left
$\frac{1}{8}$
$\frac{1}{8}$
$\frac{1}{8}$

_____ blue − _____ blue = _____ blue

	Build this tower.		Take away this many.		Difference
2.	$\frac{9}{10}$	$-$	$\frac{2}{10}$	$=$	$\frac{\square}{10}$

Take away $\frac{2}{10}$

What's left? →

_____ purple − _____ purple = _____ purple

SUBTRACTING MORE FRACTIONS

Use the grids below to find the differences. Name ______________________

1. $\frac{5}{10} - \frac{2}{10} =$ ____

2. $\frac{4}{5} - \frac{1}{5} =$ ____

3. $\frac{7}{8} - \frac{2}{8} =$ ____

4. $\frac{11}{12} - \frac{6}{12} =$ ____

5. $\frac{8}{10} - \frac{5}{10} =$ ____

6. $\frac{8}{12} - \frac{1}{12} =$ ____

Try this one.

7. $\frac{7}{8} - \frac{0}{8} =$ ____

Build first fraction

Take away

What's left?

SUBTRACTING FRACTIONS AND SIMPLIFYING DIFFERENCES

Build and draw the problem and the difference. Then build the difference with the fewest cubes. Write the difference, then simplify the difference, and write it in simplest form.

Name ____________________

1. $\frac{7}{8} - \frac{3}{8} = \frac{\square}{8} = \frac{\square}{2}$

(difference) (simplest form)

$\frac{1}{8}$ $\frac{1}{8}$ $\frac{1}{8}$ — Take away $\frac{3}{8}$

$\frac{1}{8}$ $\frac{1}{8}$ $\frac{1}{8}$ $\frac{1}{8}$ — What's left? → $\frac{1}{8}$ $\frac{1}{8}$ $\frac{1}{8}$ $\frac{1}{8}$ = $\frac{1}{2}$

_____ blue − _____ blue = _____ blue = _____ pink

2. $\frac{9}{10} - \frac{3}{10} = \frac{\square}{10} = \frac{\square}{5}$

(difference) (simplest form)

Take away $\frac{3}{10}$

What's left? → ____ = ____

_____ purple − _____ purple = _____ purple = _____ green

SUBTRACTING MORE FRACTIONS AND SIMPLIFYING DIFFERENCES

Name ______________________

Use the grids below to find the difference and then the difference in simplest form.

	difference	simplest difference
1. $\frac{4}{6} - \frac{1}{6}$ =	____ =	____
2. $\frac{7}{12} - \frac{3}{12}$ =	____ =	____
3. $\frac{5}{8} - \frac{3}{8}$ =	____ =	____
4. $\frac{9}{10} - \frac{4}{10}$ =	____ =	____
5. $\frac{11}{12} - \frac{2}{12}$ =	____ =	____
6. $\frac{5}{6} - \frac{4}{6}$ =	____ =	____

Take away

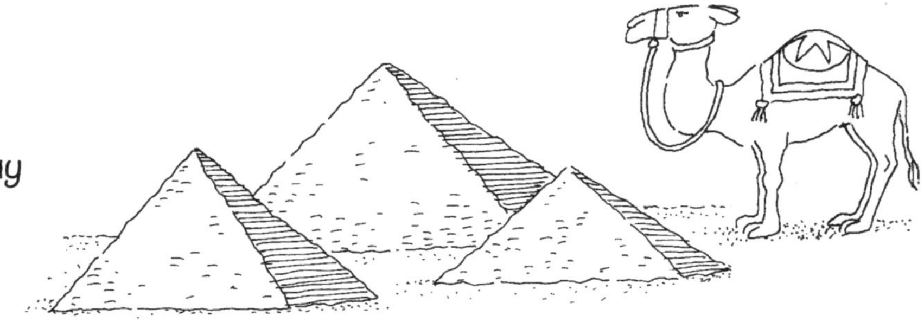

What's left?

difference

simplest difference

SUBTRACTING FRACTIONS FROM ONE

Name ______________________

1. Build fraction towers equivalent to 1. Draw the number of cubes below, and the fraction for each color.

1 red = __2__ pink = $\frac{\square}{\square}$

= _____ orange = $\frac{\square}{\square}$

= _____ yellow = $\frac{\square}{\square}$

= _____ green = $\frac{\square}{\square}$

= _____ teal = $\frac{\square}{\square}$

= _____ blue = $\frac{\square}{\square}$

= _____ purple = $\frac{\square}{\square}$

= _____ black = $\frac{\square}{\square}$

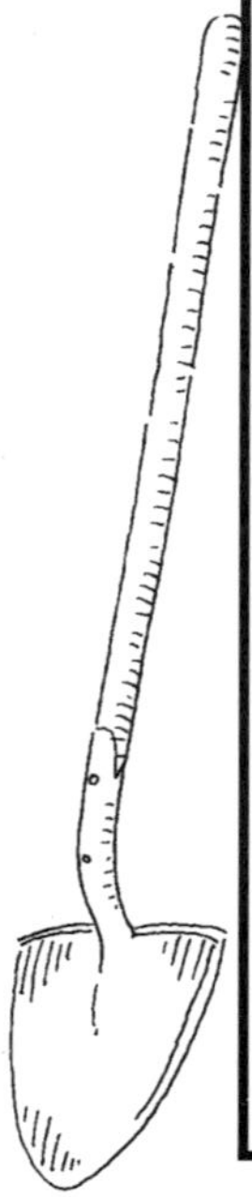

2. To subtract a fraction from 1, express 1 as a fraction with the same denominator as the fraction to be subtracted. Build and draw this problem to find the difference.

$$1 - \frac{2}{5} = \frac{\square}{5} - \frac{2}{5} = \frac{\square}{5}$$

1 → $\frac{1}{5}$ $\frac{1}{5}$ $\frac{1}{5}$ $\frac{1}{5}$ $\frac{1}{5}$

Take away $\frac{2}{5}$

What's left? → $\frac{1}{5}$ $\frac{1}{5}$ $\frac{1}{5}$

_____ red = _____ green difference

SUBTRACTING MORE FRACTIONS FROM ONE

Use the grids below to change 1 red cube to a different color. Name ______________________
Then find the difference in simplest form.

1. $1 - \frac{5}{8} =$ ____

2. $1 - \frac{8}{10} =$ ____

3. $1 - \frac{2}{6} =$ ____

4. $1 - \frac{3}{5} =$ ____

5. $1 - \frac{3}{12} =$ ____

6. $1 - \frac{6}{8} =$ ____

Try this one.

7. $2 - \frac{4}{12} =$ ____

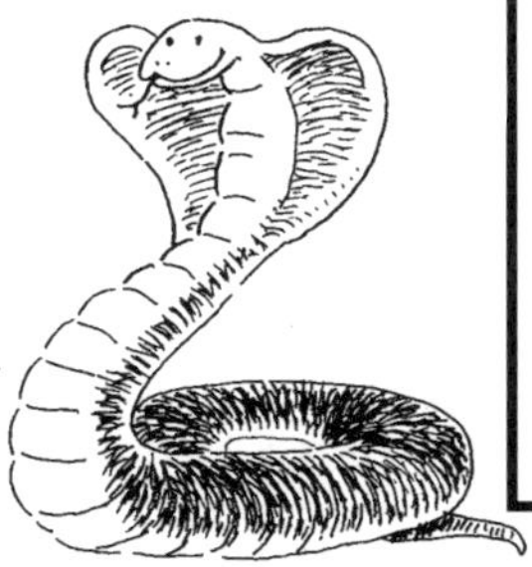

1	=		Take away	What's left? →	

$1 = \frac{\square}{\square}$ difference

SUBTRACTING FRACTIONS FROM MIXED NUMBERS

Name ______________________

Build and draw the mixed number. Then take away the fraction amount and draw the difference. Write the difference in simplest form.

1. $1\frac{5}{6} - \frac{4}{6} =$ ____

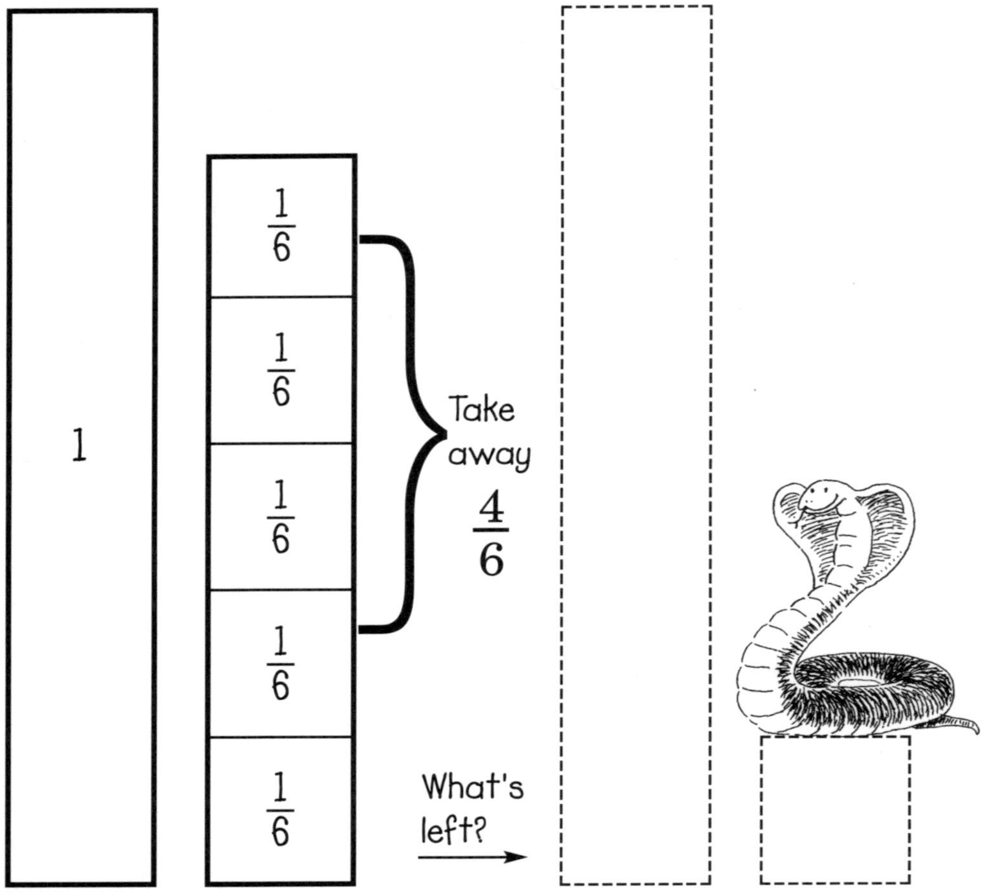

1 red, _____ teal – _____ teal = _____ red, _____ teal

2. $1\frac{7}{8} - \frac{3}{8} =$ ____

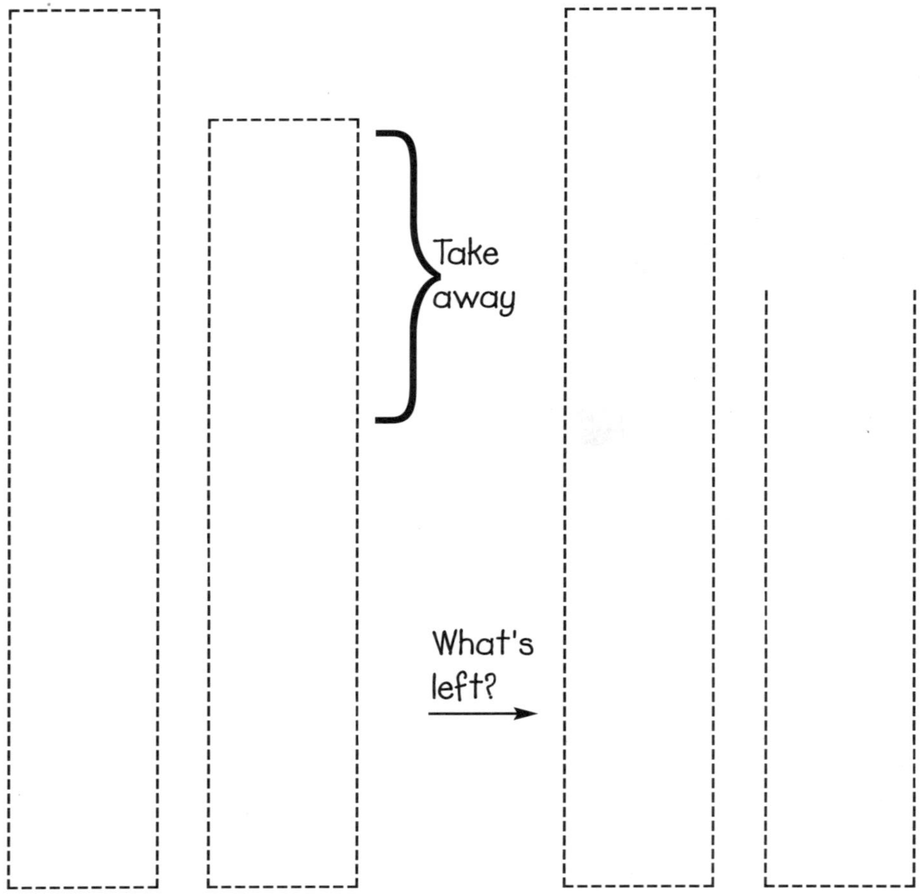

1 red, _____ blue – _____ blue = _____ red, _____ pink

SUBTRACTING MORE FRACTIONS FROM MIXED NUMBERS

Name ______________________

Use the grids below to find the differences.
Write the differences in simplest form.

1. $1\frac{7}{10} - \frac{4}{10} =$ ____

2. $1\frac{7}{8} - \frac{4}{8} =$ ____

3. $1\frac{3}{4} - \frac{1}{4} =$ ____

4. $1\frac{5}{6} - \frac{3}{6} =$ ____

5. $1\frac{11}{12} - \frac{4}{12} =$ ____

6. $1\frac{8}{12} - \frac{2}{12} =$ ____

Try this one.

7. $1\frac{3}{8} - \frac{5}{8} =$ ____

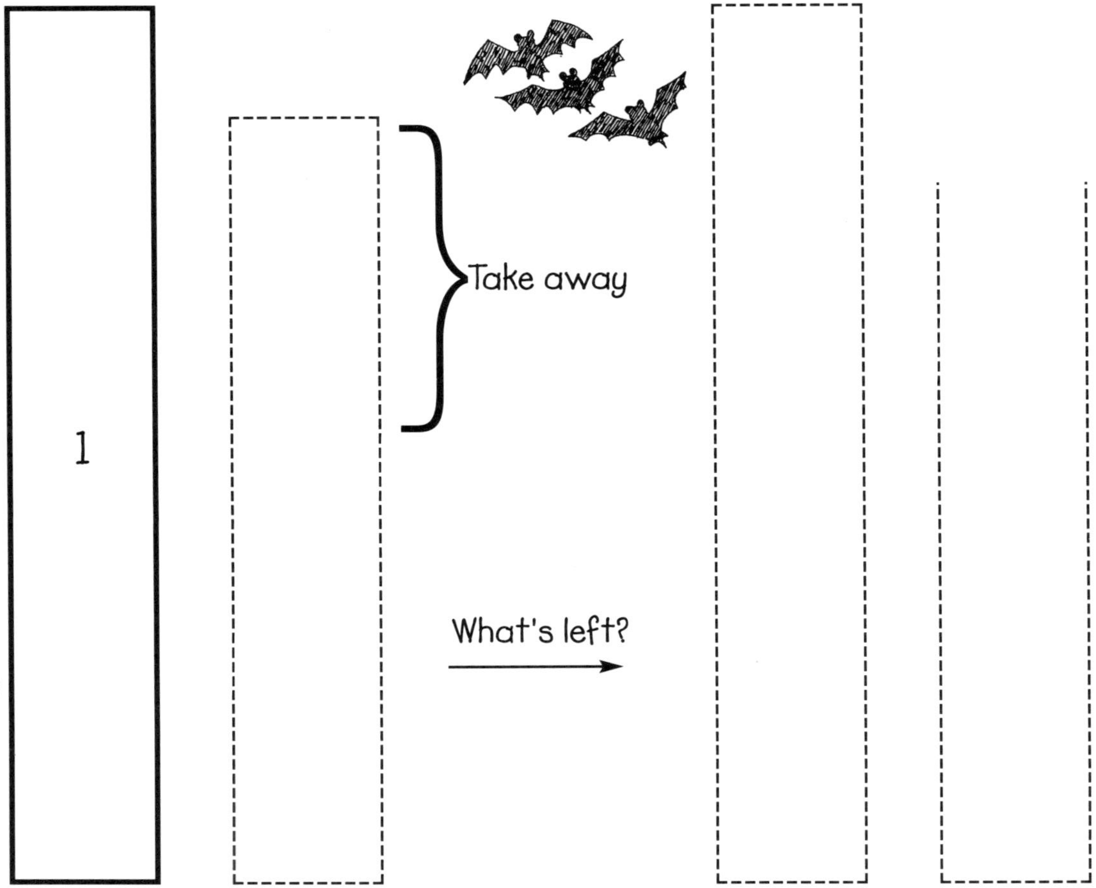

mixed number – fraction = difference

SUBTRACTING FRACTIONS WITH REGROUPING

Name ______________________

Build and draw the mixed number. Then build and draw the mixed number without using a red cube. Take away the fraction amount, and draw the difference. Write the difference in simplest form.

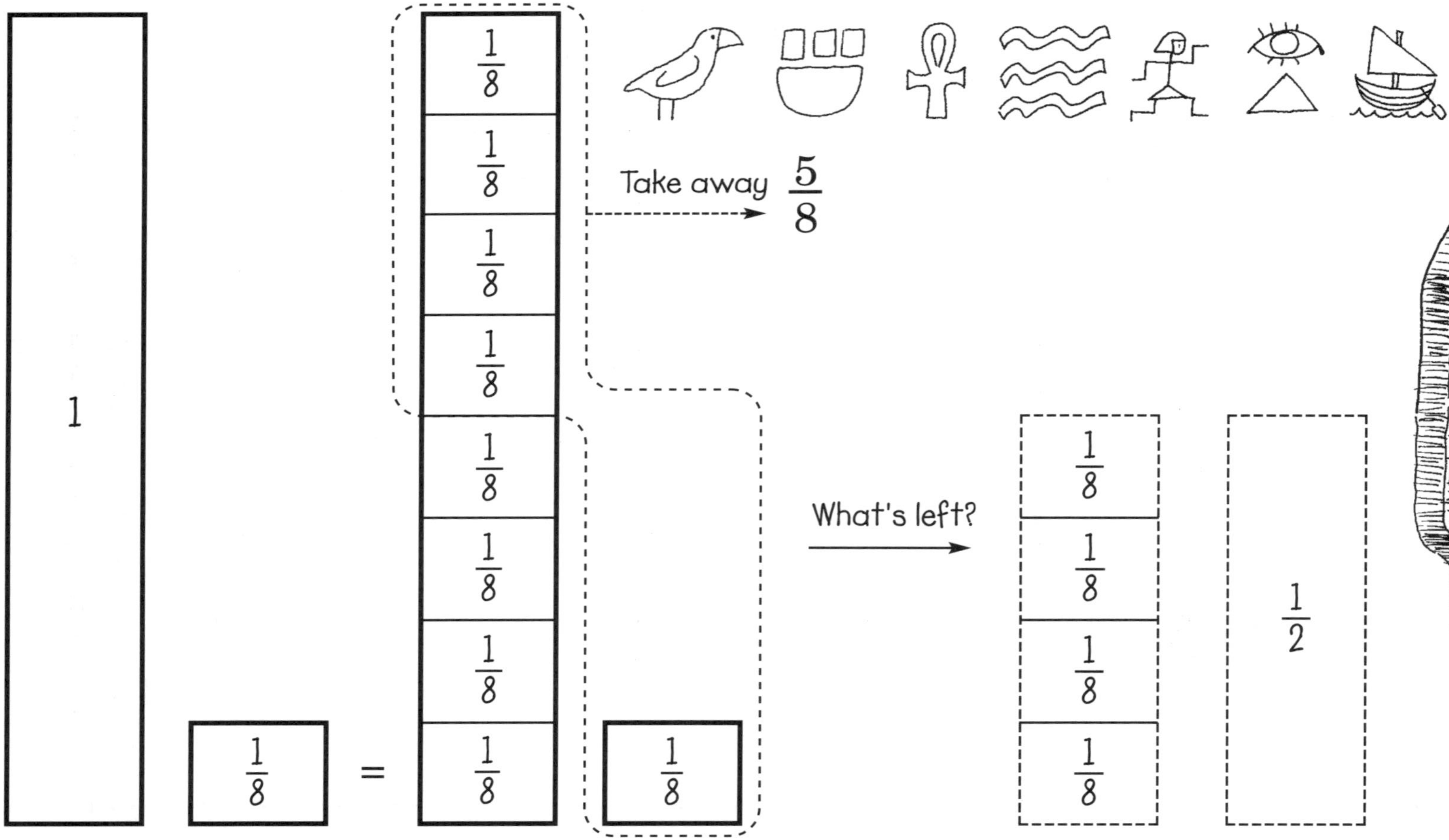

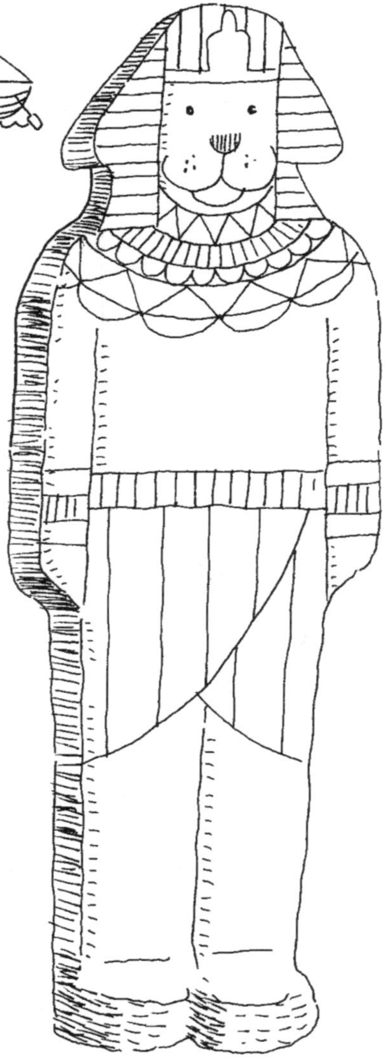

$1\frac{1}{8} = \frac{9}{8} - \frac{5}{8} = \frac{4}{8} = \frac{1}{2}$

_____ red, _____ blue = _____ blue – _____ blue = _____ blue = _____ pink

MORE SUBTRACTING WITH REGROUPING

Use the grids below to find the differences.
Write the differences in simplest form.

Name ____________________

1. $1\frac{1}{8} - \frac{3}{8} =$ ____
2. $1\frac{2}{5} - \frac{4}{5} =$ ____
3. $1\frac{5}{12} - \frac{10}{12} =$ ____
4. $1\frac{3}{10} - \frac{7}{10} =$ ____
5. $1\frac{1}{6} - \frac{2}{6} =$ ____
6. $1\frac{5}{12} - \frac{9}{12} =$ ____

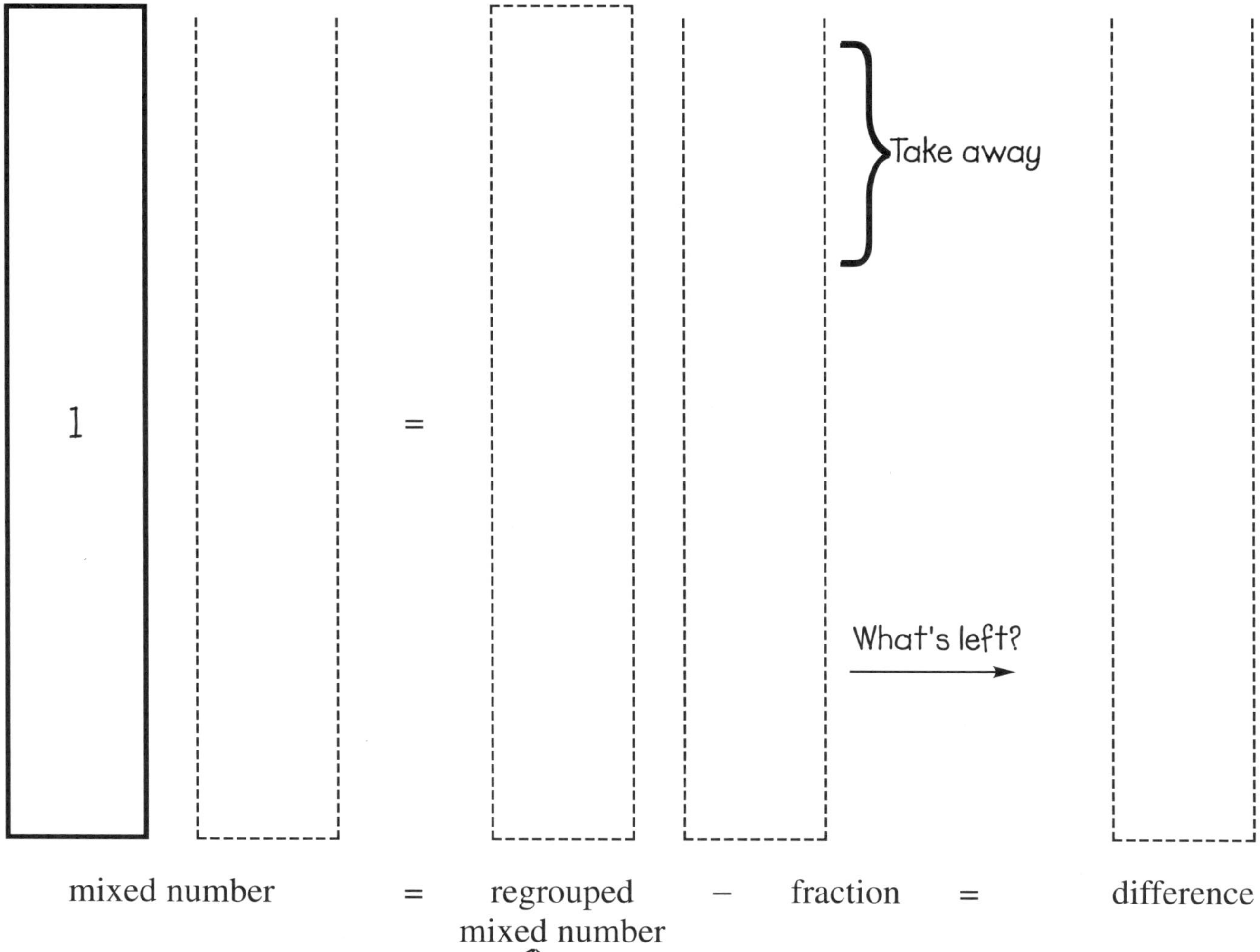

SUBTRACTING FRACTIONS AND MIXED NUMBERS

Subtract. Write the differences in simplest form. Name ______________________

1. $\frac{5}{6} - \frac{3}{6}$ = ________

2. $1\frac{7}{12} - \frac{3}{12}$ = ________

3. $\frac{7}{8} - \frac{5}{8}$ = ________

4. $\frac{9}{10} - \frac{8}{10}$ = ________

5. $1 - \frac{3}{12}$ = ________

6. $1\frac{7}{10} - \frac{9}{10}$ = ________

7. $1\frac{4}{12} - \frac{5}{12}$ = ________

ADDING AND SUBTRACTING FRACTIONS

Add or subtract. Write the answers in simplest form. Name ______________________

1. $\frac{4}{10} + \frac{1}{10} =$ ______

2. $\frac{11}{12} - \frac{4}{12} =$ ______

3. $\frac{11}{12} + \frac{4}{12} =$ ______

4. $1 - \frac{4}{10} =$ ______

5. $\frac{7}{8} + \frac{7}{8} =$ ______

6. $\frac{5}{6} - \frac{0}{6} =$ ______

7. $1\frac{1}{4} - \frac{3}{4} =$ ______

MAKE-A-ONE GAME

What will we need?

- one set of Fraction Tower® Cubes
- paper clip and pencil, or clear spinner

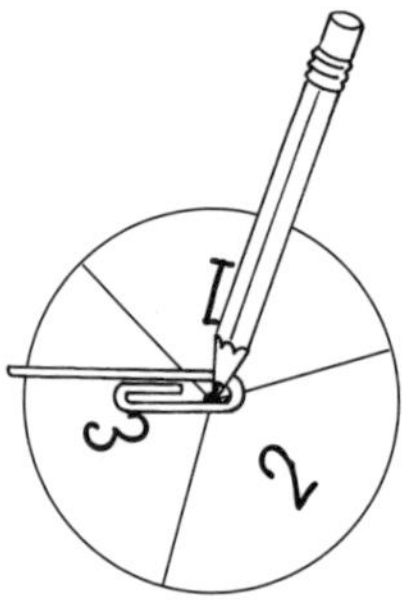

Who will play?

- 2-4 players

How will we play?

- Use only the green, blue, purple, and black fraction cubes. Place them in the middle of the table.
- Each player chooses one color: green, blue, purple, or black.
- For each turn, a player spins the spinner and takes the indicated number of fraction cubes in his or her color.
- Players keep adding fraction cubes in their color until the first one makes a one-color fraction tower for 1. The player gets 1 point, and a new round starts. If a player only needs one cube to make a 1, but gets a 2 or 3 on the spinner, he or she loses a turn.

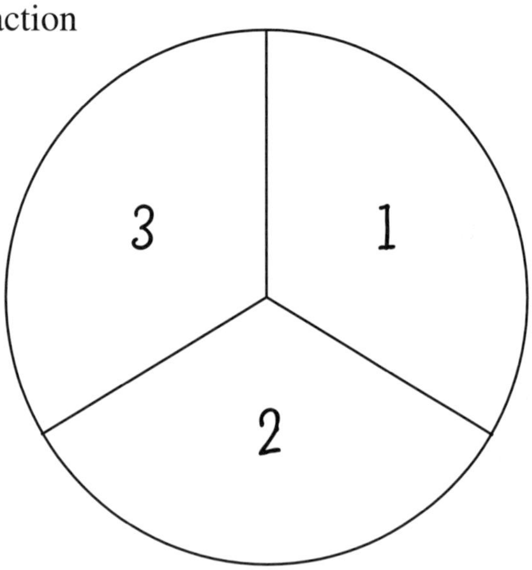

Who will win?

- The player who has the most points after 5 rounds.

SPIN-A-SUM GAME

What will we need?

- two sets of Fraction Tower® Cubes
- paper clip and pencil, or clear spinner
- 10 markers per player

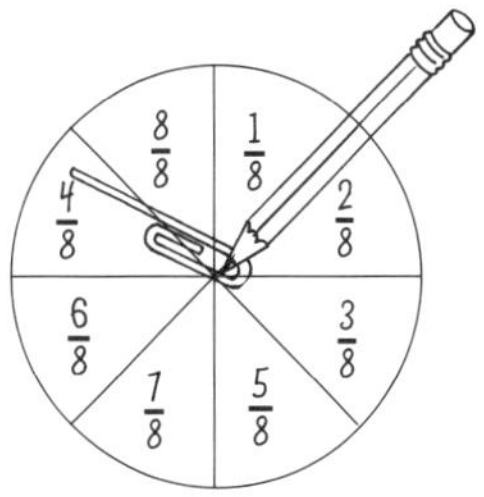

Who will play?

- two players

How will we play?

- Place the Fraction Tower Cubes in the middle of the table.
- At each turn, each player spins the spinner twice.
- The player builds the two addends with the fraction cubes, and then shows the sum in simplest form.
- Both players must agree that the sum is correct before the player can cover the sum with a marker on the Sum Gameboard. If a sum is already covered, the player loses a turn.
- Return the fraction cubes to the center of the table for the next turn.

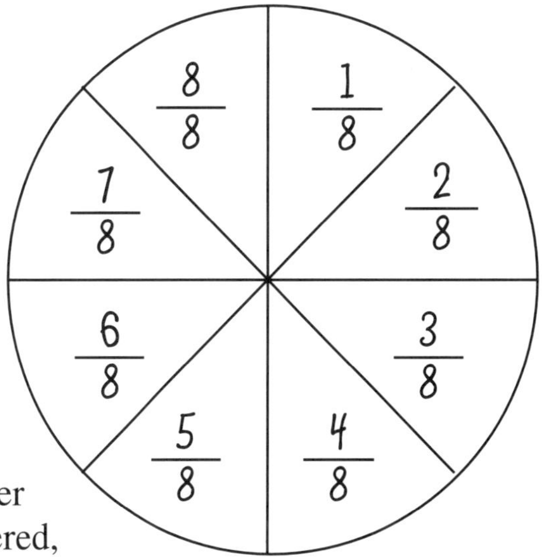

Who will win?

- The player who covers the most sums on the Sum Gameboard in ten turns.

Sum Gameboard

$\frac{1}{4}$	$\frac{3}{8}$	$\frac{1}{2}$	$\frac{5}{8}$	$\frac{3}{4}$
$1\frac{3}{8}$	$1\frac{1}{4}$	$1\frac{1}{8}$	1	$\frac{7}{8}$
$1\frac{1}{2}$	$1\frac{5}{8}$	$1\frac{3}{4}$	$1\frac{7}{8}$	2

SELECTED SOLUTIONS

Page 8: 1. $\frac{7}{10}$; $\frac{7}{10}$; 3 purple, 4 purple, 7 purple
2. $\frac{7}{8}$; $\frac{7}{8}$; 5 blue, 2 blue, 7 blue

Page 9: 1. $\frac{4}{5}$ 2. $\frac{7}{8}$
3. $\frac{5}{6}$ 4. $\frac{5}{12}$
5. $\frac{7}{10}$ 6. $\frac{10}{12}$ 7. $\frac{11}{12}$

Page 10: 1. $\frac{6}{8}$, $\frac{3}{4}$; $\frac{6}{8}$, $\frac{3}{4}$; 2 blue, 4 blue, 6 blue, 3 yellow 2. $\frac{8}{12}$, $\frac{2}{3}$; $\frac{8}{12}$, $\frac{2}{3}$; 5 black, 3 black, 8 black, 2 orange

Page 11: 1. $\frac{2}{4} = \frac{1}{2}$ 2. $\frac{3}{6} = \frac{1}{2}$
3. $\frac{6}{12} = \frac{1}{2}$ 4. $\frac{8}{10} = \frac{4}{5}$
5. $\frac{9}{12} = \frac{3}{4}$ 6. $\frac{6}{8} = \frac{3}{4}$

Page 12: $\frac{7}{6}$, $1\frac{1}{6}$; $\frac{7}{6}$, $1\frac{1}{6}$, $1\frac{1}{6}$; 3 teal, 4 teal, 7 teal, 1 red, 1 teal, 1 red, 1 teal

Page 13: 1. $\frac{5}{4} = 1\frac{1}{4}$ 2. $\frac{11}{10} = 1\frac{1}{10}$
3. $\frac{8}{5} = 1\frac{3}{5}$ 4. $\frac{17}{12} = 1\frac{5}{12}$
5. $\frac{9}{8} = 1\frac{1}{8}$ 6. $\frac{19}{12} = 1\frac{7}{12}$

Page 14: $\frac{9}{6}$, $1\frac{3}{6}$, $1\frac{1}{2}$; 5 teal, 4 teal, 9 teal, 1 red, 3 teal; 1 red, 1 pink

Page 15: 1. $\frac{12}{8} = 1\frac{4}{8} = 1\frac{1}{2}$ 2. $\frac{10}{6} = 1\frac{4}{6} = 1\frac{2}{3}$
3. $\frac{20}{12} = 1\frac{8}{12} = 1\frac{2}{3}$ 4. $\frac{12}{10} = 1\frac{2}{10} = 1\frac{1}{5}$
5. $\frac{21}{12} = 1\frac{9}{12} = 1\frac{3}{4}$

Page 16: A. Problems 2 and 4 because the sum of the numerators is less than the denominator.
B. 1. $1\frac{1}{4}$ 2. $\frac{9}{10}$
3. $1\frac{2}{3}$ 4. $\frac{2}{3}$
5. $1\frac{3}{10}$ 6. $1\frac{1}{3}$

Page 17: 1. $\frac{6}{12}$ 2. $\frac{3}{10}$
3. $\frac{7}{12}$ 4. $\frac{7}{8}$
5. $\frac{7}{12}$ 6. $\frac{5}{6}$

Page 18: 1. $\frac{3}{8}$; 7 blue, 4 blue, 3 blue
2. $\frac{7}{10}$; 9 purple, 2 purple, 7 purple

Page 19: 1. $\frac{3}{10}$ 2. $\frac{3}{5}$
3. $\frac{5}{8}$ 4. $\frac{5}{12}$
5. $\frac{3}{10}$ 6. $\frac{7}{12}$ 7. $\frac{7}{8}$

Page 20: 1. $\frac{4}{8}$, $\frac{1}{2}$; 7 blue, 3 blue, 4 blue, 1 pink
2. $\frac{6}{10}$, $\frac{3}{5}$; 9 purple, 3 purple, 6 purple, 3 green

Page 21: 1. $\frac{3}{6} = \frac{1}{2}$ 2. $\frac{4}{12} = \frac{1}{3}$
3. $\frac{2}{8} = \frac{1}{4}$ 4. $\frac{5}{10} = \frac{1}{2}$
5. $\frac{9}{12} = \frac{3}{4}$ 6. $\frac{1}{6} = \frac{1}{6}$

Page 22: 1. 2 pink = $\frac{2}{2}$; 3 orange = $\frac{3}{3}$; 4 yellow = $\frac{4}{4}$; 5 green = $\frac{5}{5}$; 6 teal = $\frac{6}{6}$; 8 blue = $\frac{8}{8}$; 10 purple = $\frac{10}{10}$; 12 black = $\frac{12}{12}$
2. $\frac{5}{5}$, $\frac{3}{5}$; 1 red, 5 green

Page 23: 1. $\frac{3}{8}$ 2. $\frac{1}{5}$
3. $\frac{2}{3}$ 4. $\frac{2}{5}$
5. $\frac{3}{4}$ 6. $\frac{1}{4}$ 7. $1\frac{2}{3}$

Page 24: 1. $1\frac{1}{6}$; 1 red, 5 teal, 4 teal; 1 red, 1 teal
2. $1\frac{1}{2}$; 1 red, 7 blue, 3 blue; (1 red, 4 blue) 1 red, 1 pink

Page 25: 1. $1\frac{3}{10}$ 2. $1\frac{3}{8}$
3. $1\frac{1}{2}$ 4. $1\frac{1}{3}$
5. $1\frac{7}{12}$ 6. $1\frac{1}{2}$ 7. $\frac{3}{4}$

Page 26: 1 red, 1 blue,
9 blue, 5 blue, 4 blue, 1 pink

Page 27: 1. $\frac{3}{4}$ 2. $\frac{3}{5}$
3. $\frac{7}{12}$ 4. $\frac{3}{5}$
5. $\frac{5}{6}$ 6. $\frac{2}{3}$

Page 28: 1. $\frac{1}{3}$ 2. $1\frac{1}{3}$
3. $\frac{1}{4}$ 4. $\frac{1}{10}$
5. $\frac{3}{4}$ 6. $\frac{4}{5}$ 7. $\frac{11}{12}$

Page 29: 1. $\frac{1}{2}$ 2. $\frac{7}{12}$
3. $1\frac{1}{4}$ 4. $\frac{3}{5}$
5. $1\frac{3}{4}$ 6. $\frac{5}{6}$ 7. $\frac{1}{2}$